ILLINOIS

IRON & BOLT COMPANY

MANUFACTURERS OF

CAST and STEEL WAGON SKEINS,

HYDRAULIC PRESSES,

BLACKSMITHS' TOOLS

VISES, ANVILS,

❧JACK SCREWS❧

SAD IRONS, CLOTHES REELS,

—AND—

LAWN ✦ ORNAMENTS.

CARPENTERSVILLE,

KANE COUNTY,   ILLINOIS.

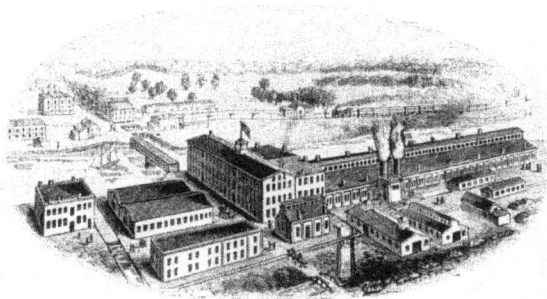

ILLINOIS IRON & BOLT COMPANY'S
WORKS AT CARPENTERSVILLE,
KANE CO., ILLS.

Astragal Press
www.astragalpress.com

Reprint © 2016
ISBN: 978-1-931626-37-8

Terms of Sale.

———◆———

All goods are delivered f. o. b. cars at Carpentersville, unless otherwise arranged.

Freight rates to any point in the United States will be furnished on application.

We secure the lowest possible rates that can be obtained, on all shipments made by us.

Ninety days credit will be given to responsible parties, or 3 per cent. discount allowed for cash if received within 15 days from date of invoice.

All payments to be made in funds current in Chicago or New York.

Carpentersville, Ills.,

June 1, 1889.

CAST IRON WAGON SKEINS.

D. V. & CO. BRAND.

PRICE LIST.

SIZE.	PER SET.	SIZE.	PER SET.	SIZE.	PER SET.
2×6 in.	$3 25	$2\frac{3}{4} \times 8$ in.	$5 25	$3\frac{1}{2} \times 11$ in.	$ 8 30
$2 \times 6\frac{1}{2}$ "	3 25	$2\frac{3}{4} \times 8\frac{1}{2}$ "	5 40	$3\frac{1}{2} \times 12$ "	8 80
$2\frac{1}{8} \times 6\frac{1}{2}$ "	3 50	$2\frac{3}{4} \times 9$ "	6 00	$3\frac{3}{4} \times 11$ "	9 00
$2\frac{1}{4} \times 7$ "	3 75	3×9 "	6 00	$3\frac{3}{4} \times 12$ "	9 40
$2\frac{1}{4} \times 7\frac{1}{2}$ "	4 00	$3\frac{1}{4} \times 9$ "	7 00	4×12 "	10 60
$2\frac{3}{8} \times 7$ "	4 00	3×10 "	6 50	$4\frac{1}{4} \times 12$ "	15 00
$2\frac{3}{8} \times 7\frac{1}{2}$ "	4 00	$3\frac{1}{4} \times 10$ "	7 50	$4\frac{1}{2} \times 12$ "	18 00
$2\frac{1}{2} \times 7$ "	4 50	$3\frac{1}{4} \times 11$ "	7 75	$4\frac{1}{2} \times 13$ "	20 00
$2\frac{1}{2} \times 7\frac{1}{2}$ "	4 50	$3\frac{1}{2} \times 10$ "	8 00	5×14 "	25 00
$2\frac{1}{2} \times 8$ "	4 75	$3\frac{1}{2} \times 10\frac{1}{2}$ "	8 00		

Discount ....

Orput Draft Lugs, as shown in Fig. 1, page 8....15c. per set extra

CAST IRON WAGON SKEINS.

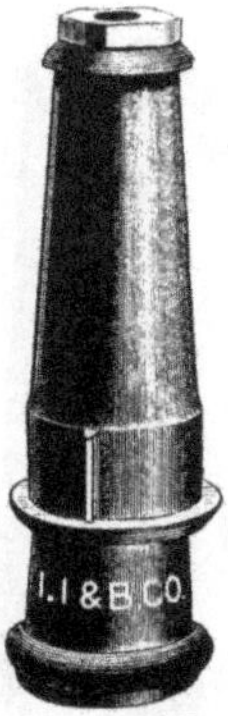

I. I. & B. BRAND.

PRICE LIST.

SIZE.	PER SET.	SIZE.	PER SET.	SIZE.	PER SET.
2 ×6½ in.	$3 25	2¾× 8½ in.	$5 40	3½×10 in.	$ 8 00
2¼×7 "	3 75	2¾× 9 "	6 00	3½×10½ "	8 00
2¼×7½ "	4 00	3 × 9 "	6 00	3½×11 "	8 30
2⅜×7½ "	4 00	3¼× 9 "	7 00	3½×12 "	8 80
2½×7½ "	4 50	3 ×10 "	6 50	3¾×11 "	9 00
2½×8 "	4 75	3¼×10 "	7 50	3¾×12 "	9 40
2¾×8 "	5 25	3¼×11 "	7 75	4½×12 "	18 00

Discount............

Oilers.................................... 20c. per set extra

Ulrich Truss Lugs, as shown in Fig. 4 on page 8, 25c. " "

CAST IRON WAGON SKEINS.

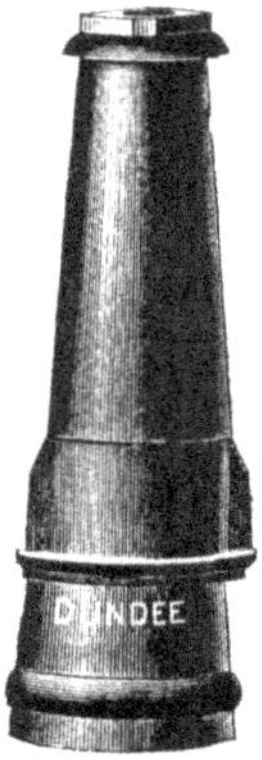

DUNDEE BRAND.

PRICE LIST.

SIZE.	PER SET.	SIZE.	PER SET.	SIZE.	PER SET.
2 ×6½ in.	$3 25	2¾× 8 in.	$5 25	3¾×11 in.	$ 9 00
2⅛×6½ "	3 50	2¾× 8½ "	5 40	3¾×12 "	9 40
2⅛×7 "	3 75	2¾× 9 "	6 00	4 ×12 "	10 60
2¼×7 "	3 75	3 × 9 "	6 00	4¼×12 "	15 00
2¼×7½ "	4 00	3¼×10 "	7 50	4½×13 "	20 00
2⅜×7½ "	4 00	3¼×11 "	7 75	5 ×13 "	23 00
2½×7 "	4 50	3½×10 "	8 00	5 ×14 "	25 00
2½×7½ "	4 50	3½×11 "	8 30	5¼×14 "	27 50
2½×8 "	4 75	3½×12 "	8 80		

Discount............

SPECIAL CAST IRON WAGON SKEINS.

EXTRA HEAVY.

No. 1.   No. 2.   No. 3.   No. 4.

No. 1. Shows the Orput patent Lug used on front axles for attaching draft or stay chains. Can furnish $2\frac{1}{2}$x8, $2\frac{3}{4}$x8, 3x9, $3\frac{1}{4}$x10, $3\frac{1}{2}$x11 and $3\frac{3}{4}$x12 of this style heavy pattern. The boxes of these Skeins are made to wedge in.

No. 2. Shows a Lug on under side of Skein which is intended to receive a round Truss-rod. Can furnish $2\frac{1}{2}$x$7\frac{1}{2}$, $2\frac{3}{4}$x$8\frac{1}{2}$, 3x9, $3\frac{1}{4}$x10 and $3\frac{1}{2}$x11, with boxes made to wedge in.

No. 3. Shows a Skein with another style of Lug for attaching Truss-rod. Can furnish $2\frac{1}{2}$x$7\frac{1}{2}$, $2\frac{3}{4}$x$8\frac{1}{2}$, 3x9, $3\frac{1}{4}$x10 and $3\frac{1}{2}$x11, with boxes made to wedge in.

No. 4. Shows Ulrich's patent Truss attachment. Can furnish all sizes of this style, with boxes made to wedge in.

———

The above Skeins are heavier than the regular Trade Skeins. Net prices will be furnished on application.

SPECIAL CAST IRON WAGON SKEINS.

EXTRA HEAVY

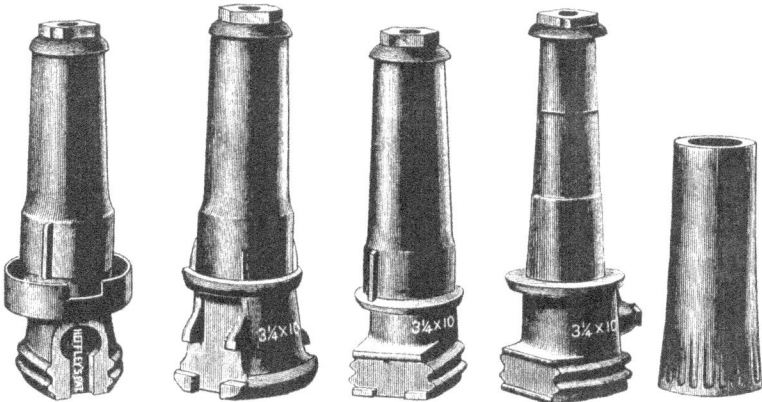

No. 5.　　　　No. 6.　　　　No. 7.　　　　No. 8.

No. 5.　Represents Heffley's patent Skein with sand bands, and shows device for attaching Clip and Truss-bar. These skeins are made with boxes to wedge in. A more descriptive circular, with a list of sizes that we can furnish, will be mailed on application.

No. 6.　Shows a Skein made to receive Clip and Truss-bar. The Clip-bar is intended to be welded to the Truss-bar in the form of a **T.** Can furnish $2\frac{1}{4}$x$7\frac{1}{2}$, $2\frac{1}{2}$x8, $2\frac{3}{4}$x$8\frac{1}{2}$, 3x9, $3\frac{1}{4}$x10, $3\frac{1}{2}$x11 and $3\frac{3}{4}$x12 of this style, with or without Oilers, and with boxes made to press in.

No. 7.　Shows another Skein made to receive Clip and Truss-bar, the Clip-bar to be welded to the Truss-bar in the shape of a **T.** Can furnish $2\frac{3}{4}$x$8\frac{1}{2}$, 3x9, $3\frac{1}{4}$x10 and $3\frac{1}{2}$x11, with either press-in or wedge-in boxes, and with or without Oilers.

No. 8.　Shows an oiler Skein with chambered arm and a fluted Box to press in, also device for clipping. Can furnish $2\frac{1}{4}$x8, $2\frac{3}{4}$x8, $2\frac{3}{4}$x$8\frac{1}{2}$, 3x9, $3\frac{1}{4}$x10, $3\frac{1}{2}$x11, $3\frac{3}{4}$x12 and 4x12, with or without Oilers.

The above Skeins are heavier than the regular Trade Skeins. Net prices will be furnished on application.

STANDARD STEEL WAGON SKEINS.

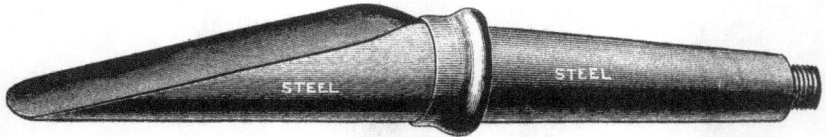

These Skeins are thoroughly welded, smooth and uniform in size, and only the best material is used in their manufacture.

PRICE LIST.

Size.	Gauge of Steel.	Price per set.	Size.	Gauge of Steel.	Price per set.	Size.	Gauge of Steel.	Price per set.
2¼×7	No. 11	$5 60	3 ×10	No. 10	$6 70	3½×12	No. 9	$ 9 00
2¼×7½	" 11	5 60	3¼×9	" 9	7 40	3¾×11	" 8	10 00
2½×7½	" 11	5 70	3¼×10	" 9	7 50	3¾×12	" 8	10 00
2½×8	" 11	5 80	3¼×11	" 9	7 60	4 ×12	" 7	12 00
2¾×8	" 10	6 10	3½×10	" 9	8 30	4¼×12	" 7	13 00
2¾×8½	" 10	6 20	3½×10½	" 9	8 40	4½×12	" 6	17 00
2¾×9	" 10	6 30	3½×11	" 9	8 50	4½×13	" 6	18 00
3 ×9	" 10	6 50						

Discount...........

I. I. & B. CO.'S STEEL WAGON SKEINS.

PATENTED MAY 15, 1883.

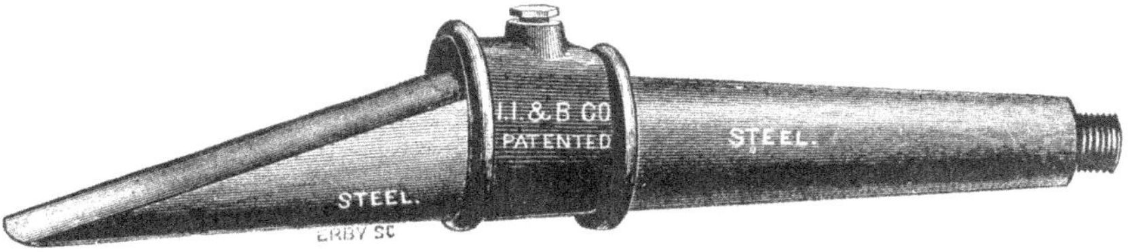

Our patent Steel Skeins are giving general satisfaction. Their great strength and symmetrical appearance recommend them at once to the buyer of wagons.

The sleeves of these Skeins having the same shape, uniformity and smoothness of cast iron skeins, they can be fitted to the axletree with the same machine as is used in fitting the cast skeins, and with no additional labor.

We can furnish these Skeins with Draft or Truss Lugs and Oilers, the same as on cast Skeins.

PRICE LIST.

Size.	Gauge of Steel.	Price per set.	Size.	Gauge of Steel.	Price per set.	Size.	Gauge of Steel.	Price per set.
2¼×7	No. 11	$5 60	3 ×10	No. 10	$6 70	3½×12	No. 9	$ 9 00
2¼×7½	" 11	5 60	3¼×9	" 9	7 40	3¾×11	" 8	10 00
2½×7½	" 11	5 70	3¼×10	" 9	7 50	3¾×12	" 8	10 00
2½×8	" 11	5 80	3¼×11	" 9	7 60	4 ×12	" 7	12 00
2¾×8	" 10	6 10	3½×10	" 9	8 30	4¼×12	" 7	13 00
2¾×8½	" 10	6 20	3½×10½	" 9	8 40	4½×12	" 6	17 00
2¾×9	" 10	6 30	3½×11	" 9	8 50	4½×13	" 6	18 00
3 ×9	" 10	6 50						

Discount............

Self Oilers................................ 25c. per set extra
Orput Draft Lugs........................ 15c. " "
Ulrich Truss Lugs....................... 25c. " "

HYDRAULIC PRESS.
NO. 1.

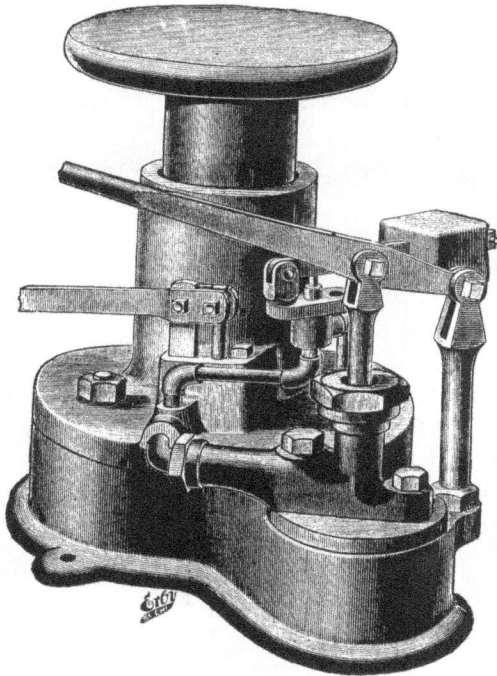

This Press has a capacity of 12 tons pressure.

The pump is made of brass, the balance of cast and wrought iron.

Diameter of ram plunger, 4 inches.

Diameter of pump plunger, $\frac{3}{4}$ inch.

It will raise 4 inches. Weight, 180 pounds.

It can be operated by hand or power.

Manufacturers of light wagons and carriages will find it very convenient for pressing on skeins and hub bands, and for pressing in boxes. It can also be used to advantage for other purposes where pressure is required.

Full directions are furnished with each machine.

Price, boxed...$50 00

Discount...........,

HYDRAULIC PRESS.

NO. 2.

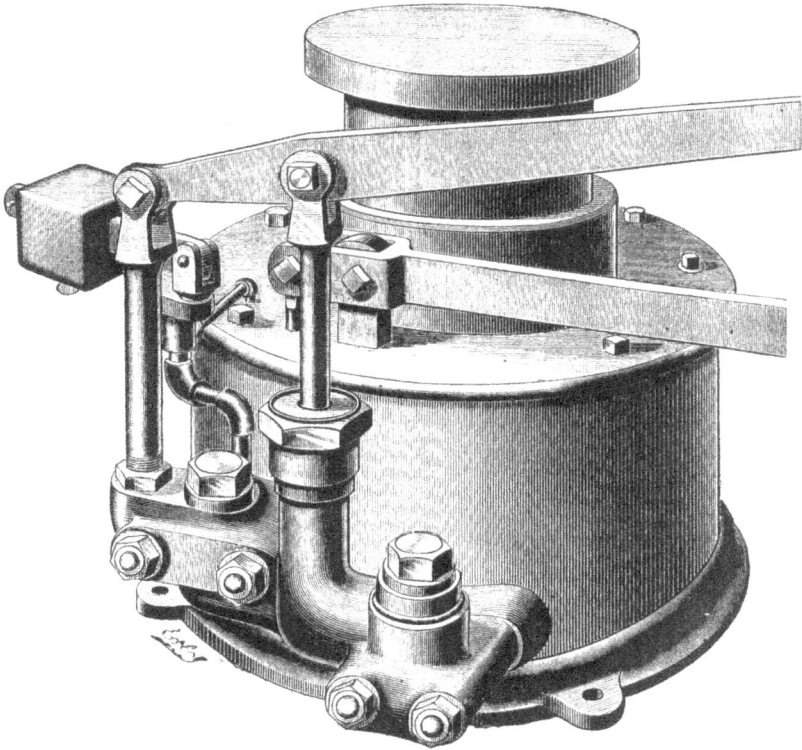

This Press has a capacity of 50 tons pressure. It is intended for pressing boxes into hubs, and for pressing on hub bands for farm and freight wagons, and other purposes where great pressure is required.

The pump is made of brass, the balance of cast and wrought iron.

Diameter of ram plunger, 8 inches.

Diameter of pump plunger, from 1 to $1\frac{1}{4}$ inches.

It will raise 6 inches. Weight, 600 pounds.

It can be operated by hand or power.

Full directions are furnished with each machine.

Price, boxed...$110 00

Discount............

SWAGE BLOCKS.

CAST IRON.

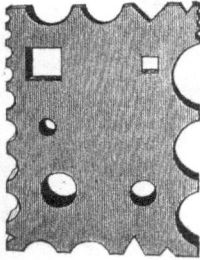

Shape of Nos. 1, 2 and 3.   Shape of Nos. 4 and 5.

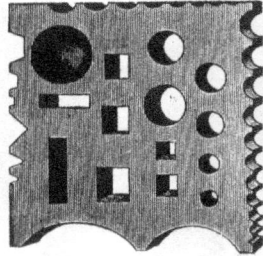

No. 1.   Measures 3⅝x10x14 inches..........Weight about 100 lbs.
" 2.   "   3⅞x11x15   "   .... "   "   125 "
" 3.   "   4¼x11x15   "   .... "   "   145 "
" 4.   "   4 x15x15   "   .... "   "   165 "
" 5.   "   6 x24x24   "   .... "   "   625 "

Price per lb.....................................

B. S. CONES OR MANDRELS.

CAST IRON.

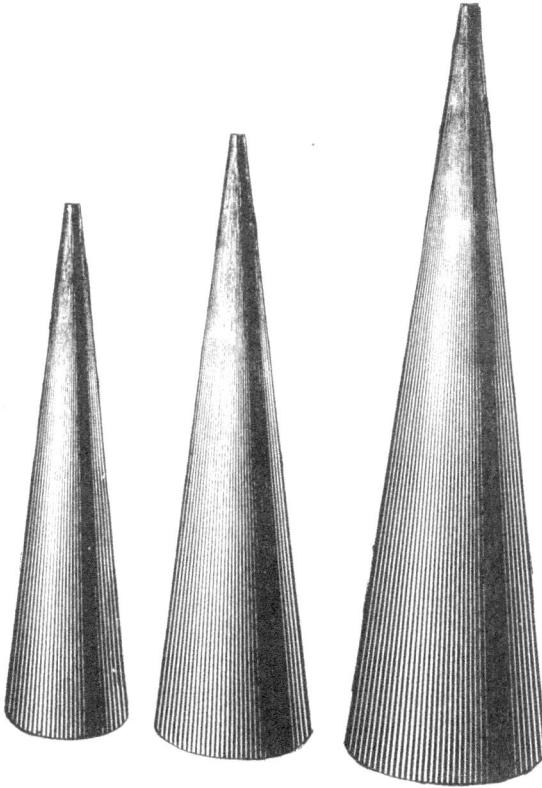

Size.	Height.	Diam. at Base.	Diam. at Top.	Weight.
No. 1	32 inches.	8 inches.	1 inch.	About 55 pounds.
" 1½	40 "	10 "	1 "	" 90 "
" 2	48 "	12 "	1 "	" 115 "
" 3	52 "	14 "	1 "	" 140 "
" 4	54 "	16 "	2 inches.	" 200 "

Price per lb......................................

CAST BOLSTER PLATES.

COMMON.

No. 1................Length, 9½ inches.......Weight, 10¾ lbs.
" 2................ " 11½ " .... " 13 "

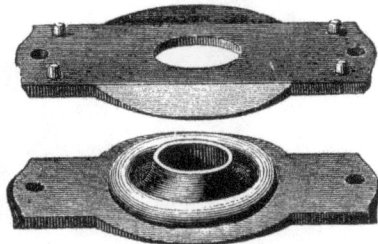

IMPROVED.

No. 1................Length, 8 inches........ Weight, 5½ lbs.
" 2................ " 10 " .... " 8½ "
" 3................ " 10¾ " .... " 11½ "

CAST REACH PLATES.

With flanges for Reaches 2¾, 3, 3¼, 3½, 3¾ and 4 inches wide.

CAST WAGON RUB IRONS.

Price per lb..

HORIZONTAL BLACKSMITH DRILLS.

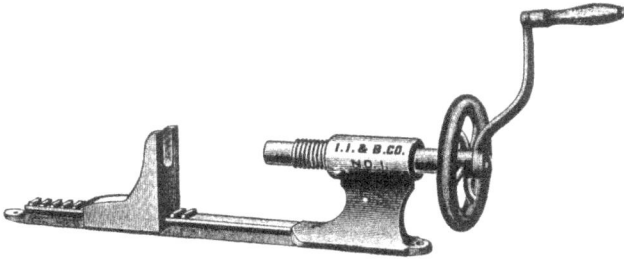

No. 1.   Solid Standard................................Price, $2 40

No. 2...Price, $2 50

"   3.   Heavy pattern...........................   "   2 70

Discount............

ON the following pages we illustrate our line of Upright Drills, and invite the attention of all who are interested in the sale and use of such machines.

In getting out these Drills it has been our purpose to manufacture a line of machines that are strong and durable, easy running, will do first-class work and can be sold at a moderate price.

By an adjustable device on the Bailey Drills the self-feed can be easily changed to slow or fast feed, to suit the work.

The tables are held firmly in position at any point on the column by an improved cam, easily operated.

The spindles are made of polished steel, with a nut at the upper end which takes up any wear of the spindle.

Our Bailey Drills are made with self feed, which may be used or not as desired ; the Illinois Upright Drill is made to feed by hand only.

The balance wheels can be easily removed if desired.

We use only the best material in these machines, and their construction and finish cannot be excelled.

Every Drill is set in perfect line and tested before it leaves our works.

Full directions accompany each Drill.

ILLINOIS UPRIGHT DRILL.

SIMPLE, STRONG AND DURABLE.

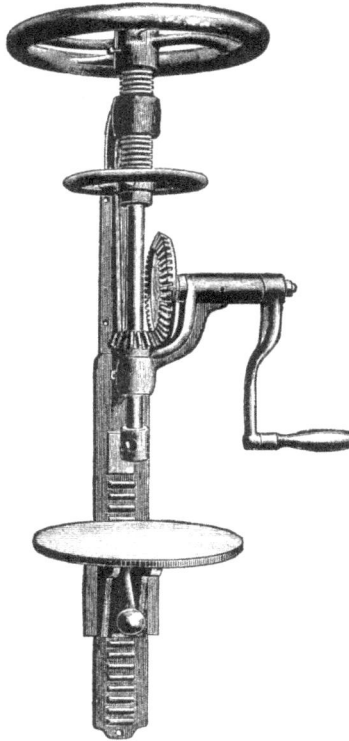

The spindle is made to receive 41-64 inch straight shank drill bits, unless otherwise ordered.

This machine will drill from 0 to ⅝-inch hole, and to the centre of a 13-inch circle.

The spindle travels 4 inches.

Price, boxed $13 00

Discount............

BAILEY DRILL.

NO. 3.

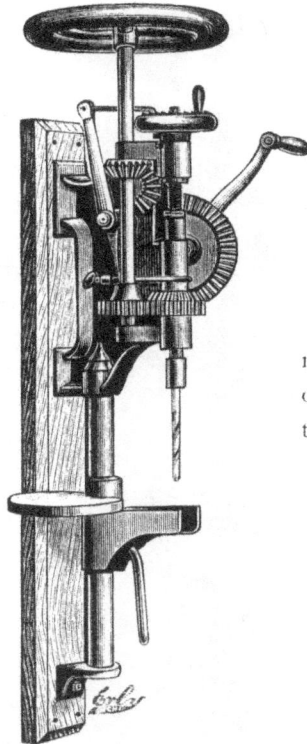

The balance-wheel shaft runs on a vertical bearing causing this Drill to run easy.

Different speeds may be obtained by changing the crank to different shafts.

The spindle of this Drill is made to receive $\frac{1}{2}$-inch straight shank drill bits, unless otherwise ordered.

This machine will drill from 0 to 1-inch hole, and to the centre of a 10-inch circle.

The spindle will travel $3\frac{1}{2}$ inches.

Price, boxed $22 00

Discount...........

BAILEY DRILL.

NO. 4.

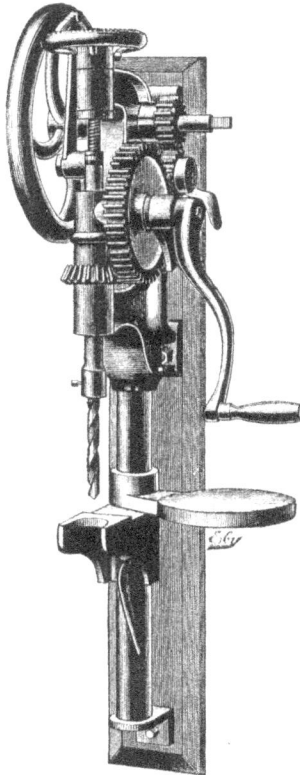

Through an inter-mediate gear the balance wheel may be easily discon-nected.

By changing the crank from one shaft to another, dif-ferent speeds may be obtained.

The spindle of this Drill is made to receive ½-inch straight shank drill bits, unless otherwise ordered.

This machine will drill from 0 to 1-inch hole, and to the center of a 10-inch circle.

The spindle will travel 3½ inches.

Price, boxed..$24 00

Discount............

22

BAILEY DRILL.

NO. 5.

The table of this Drill may be fastened at any point on the column, and can be adjusted to any desired angle convenient to the operator. The work can be clamped to the table the same as on power Drills.

Fast or slow speed may be obtained by sliding the large spur gear backward or forward.

The spindle runs in a sleeve bearing 9 inches long, which gives it great steadiness and prevents the spindle from being worn and becoming loose.

The spindle is fed up or down by a worm working in a rack upon the spindle sleeve, and is easily operated by the feed wheel.

The spindle of this Drill is made to receive $\frac{41}{64}$-inch straight shank drill bits, unless otherwise ordered.

This machine will drill from 0 to $1\frac{1}{2}$ inches, and to the center of a 16-inch circle. The spindle travels 4 inches.

Price, as shown in cut.........................Boxed, $36 00
" with loose and tight pulleys, for power only... " 40 00
" " " " " " hand or power, " 42 00
" with countershaft and cone pulleys, for power only, " 60 00

Discount............

TWIST DRILL BITS,

WITH 1-2 INCH SHANK,

FOR

NOS. 3 AND 4 BAILEY DRILLS.

Size.	Price each.	Size.	Price each.	Size.	Price each.	Size.	Price each.
$\frac{1}{8}$	$0 45	$\frac{3}{8}$	$0 80	$\frac{5}{8}$	$1 40	$\frac{27}{32}$	$2 30
$\frac{5}{32}$	45	$\frac{13}{32}$	85	$\frac{21}{32}$	1 50	$\frac{7}{8}$	2 45
$\frac{3}{16}$	50	$\frac{7}{16}$	90	$\frac{11}{16}$	1 60	$\frac{29}{32}$	2 60
$\frac{7}{32}$	55	$\frac{15}{32}$	95	$\frac{23}{32}$	1 70	$\frac{15}{16}$	2 75
$\frac{1}{4}$	60	$\frac{1}{2}$	1 00	$\frac{3}{4}$	1 85	$\frac{31}{32}$	2 90
$\frac{9}{32}$	65	$\frac{17}{32}$	1 10	$\frac{25}{32}$	2 00	1	3 00
$\frac{5}{16}$	70	$\frac{9}{16}$	1 20	$\frac{13}{16}$	2 15	$1\frac{1}{16}$	3 40
$\frac{11}{32}$	75	$\frac{19}{32}$	1 30				

WITH 41-64 INCH SHANK,

FOR

NO. 5 BAILEY AND ILLINOIS UPRIGHT DRILLS.

Size.	Price each.	Size.	Price each.	Size.	Price each.	Size.	Price each.
$\frac{1}{8}$	$0 55	$\frac{3}{8}$	$0 85	$\frac{5}{8}$	$1 05	$\frac{7}{8}$	$1 45
$\frac{5}{32}$	58	$\frac{13}{32}$	88	$\frac{21}{32}$	1 10	$\frac{29}{32}$	1 55
$\frac{3}{16}$	60	$\frac{7}{16}$	90	$\frac{11}{16}$	1 15	$\frac{15}{16}$	1 60
$\frac{7}{32}$	65	$\frac{15}{32}$	93	$\frac{23}{32}$	1 20	$\frac{31}{32}$	1 70
$\frac{1}{4}$	70	$\frac{1}{2}$	95	$\frac{3}{4}$	1 25	1	1 80
$\frac{9}{32}$	73	$\frac{17}{32}$	98	$\frac{25}{32}$	1 30	$1\frac{3}{32}$	1 90
$\frac{5}{16}$	75	$\frac{9}{16}$	1 00	$\frac{13}{16}$	1 35	$1\frac{1}{16}$	2 00
$\frac{11}{32}$	80	$\frac{19}{32}$	1 03	$\frac{27}{32}$	1 40	$1\frac{1}{8}$	2 20

Discount............

TIRE BENDERS.

NO. 1.

This machine is made with turned rollers and bearings, and will bend 3¼-inch tire or smaller to a circle 30 inches diameter or larger.

Price, with crank $6 00

NOS. 2 AND 2 1-2.

No. 2 is made with gear and pinion, has turned rollers and bearings, and will bend 3¼-inch tire or smaller to a circle 30 inches diameter or larger.

Price, with crank.................. ....$7 00

No. 2½ is made as No. 2, but will bend 6-inch tire or smaller to a circle 30 inches diameter or larger.

Price, with crank.......................................$8 50

Discount............

TIRE BENDER.

NO. 3.

EXTRA HEAVY.

This machine is double geared, has turned Rollers and Bearings, is very strong and durable, and will bend 5-inch Tire or smaller to a circle 30 inches diameter or larger.

In bending very heavy Tire two cranks may be used on opposite sides, thus giving double power.

Price, with one crank.................................... $10 50

Discount.............

H. & B. TIRE SHRINKER.

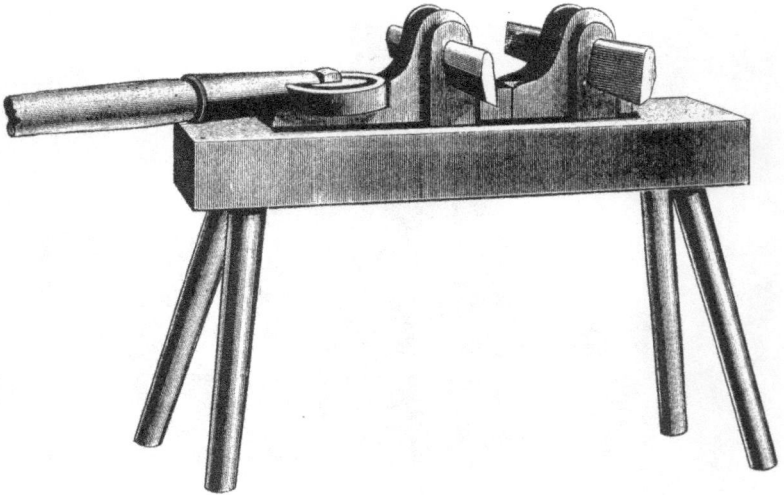

This machine will shrink Wagon Tire $2\frac{3}{4}$ inches wide or smaller, and is considered a good machine for ordinary work.

Price, with cast iron wedges................. ....$ 9 00

    "   "   steel   " 13 00

Discount............

MAGIC TIRE SHRINKER.

THE BLACKSMITH'S FAVORITE.

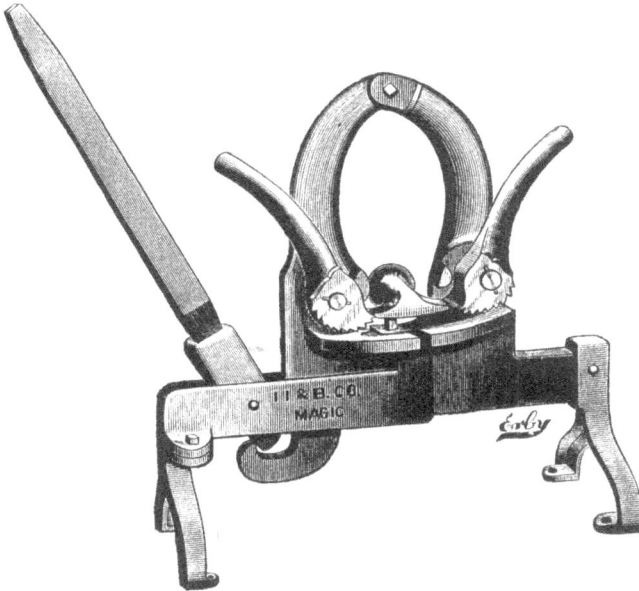

This is a strong and durable machine. It is always ready for use, and occupies but little room.

It will shrink Tire 3 inches wide or smaller very rapidly.

Price............. ....$12 00

Discount............

THE O. & D. TIRE SHRINKER AND PUNCH.

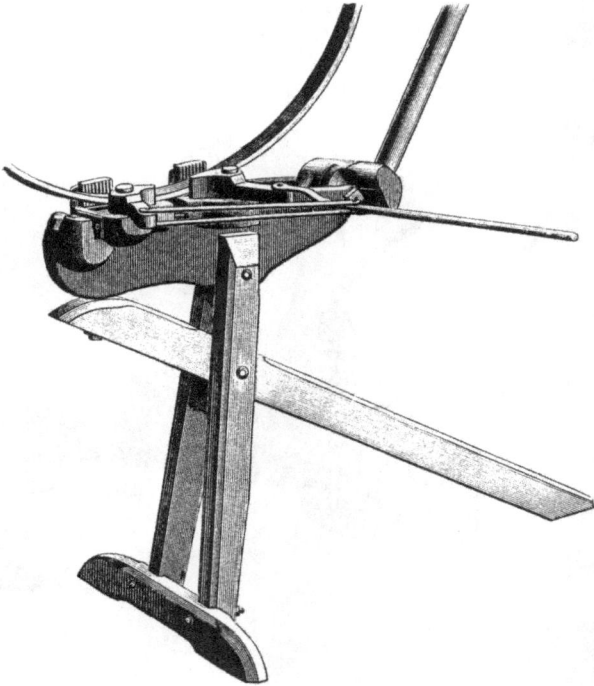

This machine has acquired the reputation of being an excellent Tire Shrinker and Punch.

It will shrink Tire $3\frac{1}{4}$ inches wide or smaller.

Price $14 00

Discount...........

I. I. & B. TIRE SHRINKER.

No. 3.

PATENTED.

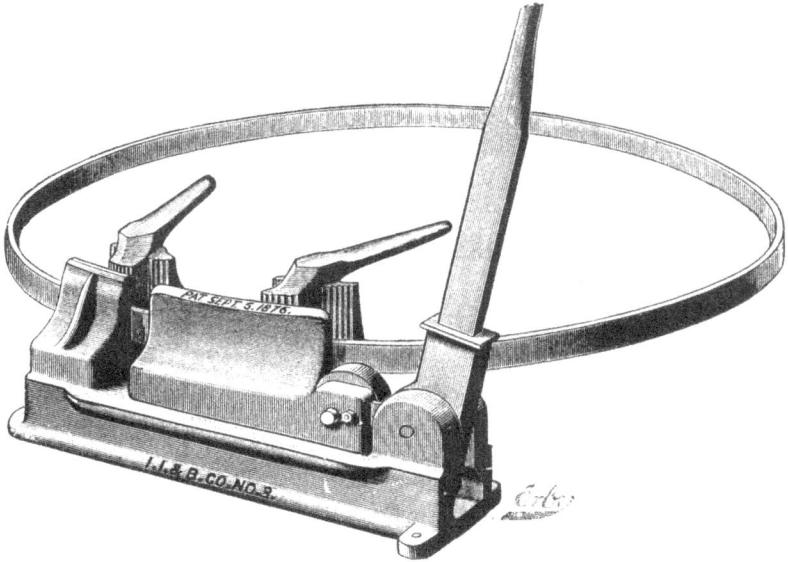

This Tire Shrinker is especially intended for heavy work, and will shrink Tire 4 inches wide or smaller.

All parts are made very strong, and will withstand a great amount of wear.

Price $16 00

Discount . . . .

TUYER IRONS.

GLOBE.

Weight without Plate, 24 lbs.............Price, $13 00 per dozen
Weight of Plate,   26 " ....   "   14 00   "

BOX.

No. 1. Weight, 16 lbs..................Price, $ 9 00 per dozen
 " 2.   "   21 " ....   "   12 00   "
 " 3.   "   25 " ....   "   14 00   "

DUCK NEST.

No. 1. Single. Weight, 11 lbs..........Price, $ 8 00 per dozen
 " 2.   "   "   15 " ....   "   10 00   "
 " 3.   "   "   18 " ....   "   12 00   "
 " 2. Double.   "   17 " ....   "   12 00   "

Discount............

TUYER IRONS.

NORTON'S PATENT.

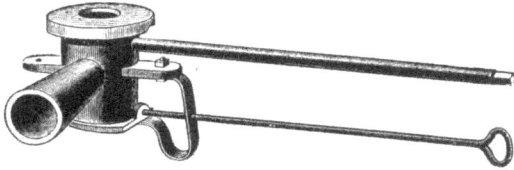

AN EXCELLENT TUYER.

To regulate the blast, turn the large rod.

The cinders and ashes may be removed by drawing out the small rod which opens the slide.

The levers and spring are readily changed to either side for right or left hand use.

Weight about 27 lbs...................Price, $24 00 per dozen

WARREN'S PATENT.

VERY SIMPLE AND EASILY ADJUSTED.

The blast is regulated by simply revolving the ball, which has three unequal sides. Open the bottom valve, and all cinders and ashes drop out.

Weight about 31 lbs...................Price, $26 00 per dozen

Discount....... ....

VISES.

The jaws of our Vises are steel faced. The beams are planed true on all sides. The screws are cut with a deep square thread, and the heads of the screws and the levers are turned and polished. A nut and washer attached to the screw at the extreme end of the beam take up the wear on the screw, and thus prevent any lost motion, so that the jaw will respond instantly to every movement of the screw. They are also provided with a lug or stop by which the nut is held firmly in position and cannot become loose.

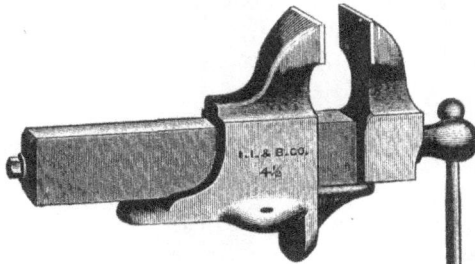

WOOD WORKER'S.
Length of Jaws, 4½ inches.   Opens 9 inches.
Price, $8 00.

MACHINIST'S.

Length of Jaws,			Opens				Price, $	
"	"	2½ inches.	"	2½ inches	"	....	"	5 50
"	"	3 "	"	3½ "	"	....	"	6 50
"	"	3½ "	"	4¼ "	5......	"	7 50
"	"	4 "	"	4½ "	"	....	"	8 50
"	"	4½ "	"	5 "	"	....	"	9 50
"	"	5 "	"	7 "	"	....	"	10 50
"	"	6 "	"	8 "	"	....	'	20 00

Discount..........

VULCAN ANVIL.

The face of this Anvil is one solid piece of tool steel, thoroughly welded to the body of the Anvil by a patented process. It is then accurately ground and tempered.

The horn is covered with and its extremity is made entirely of tough, untempered cast steel.

The body of the Anvil is made of the best Lake Superior charcoal iron, and being much more solid than wrought iron, the work being forged receives the full force of the blow. The face and horn of the Anvil are warranted to be thoroughly welded and not to separate.

PRICE LIST.

No. 00,	Weighing about 5 lbs.			$2 00	No. 8,	Weighing about 80 lbs.		$ 8 00
No. 0,	``	``	10	``	2 75	No. 9,	``	`` 90 ``	9 00
No. 1,	``	``	15	``	3 25	No. 10,	``	`` 100 ``	10 00
No. 2,	``	``	20	``	4 00	No. 11,	``	`` 110 ``	11 00
No. 3,	``	``	30	``	4 50	No. 12,	``	`` 120 ``	12 00
No. 4,	``	``	40	``	5 25	No. 13,	``	`` 130 ``	13 00
No. 5,	``	``	50	``	6 00	No. 14,	``	`` 140 ``	14 00
No. 6,	``	``	60	``	6 50	No. 15,	``	`` 150 ``	15 00
No. 7,	``	``	70	``	7 25				

Discount............

LOCOMOTIVE JACK SCREWS.

We exercise the utmost care in selecting the stock for our various styles of Jack Screws. This, together with our improved machinery, enables us to furnish an excellent line of goods at low prices.

Great strength and economy in weight are combined in the peculiar bell shape of our Jack Screw Stands.

All Screws are warranted to raise the weights specified in the following lists.

PRICE LIST

OF

LOCOMOTIVE JACK SCREWS.

Diam. of Screw.	Height of Stand.	Height Over All.	Lifting Capacity.	Price.	Diam. of Screw.	Height of Stand.	Height Over All.	Lifting Capacity.	Price.
1⅛ in.	6 in.	9 in.	8 Tons	$2 90	2 in.	18 in.	22½ in.	20 Tons	$10 25
1⅛ "	8 "	11 "	8 "	3 25	2 "	20 "	24½ "	20 "	11 50
1⅛ "	10 "	13 "	8 "	3 60	2 "	22 "	26½ "	20 "	12 50
1⅛ "	12 "	15 "	8 "	4 00	2 "	24 "	28½ "	20 "	13 50
1⅛ "	14 "	17 "	8 "	4 40	2¼ "	8 "	13 "	24 "	7 50
1¼ "	6 "	9 "	10 "	3 10	2¼ "	10 "	15 "	24 "	8 25
1¼ "	8 "	11 "	10 "	3 40	2¼ "	12 "	17 "	24 "	9 00
1¼ "	10 "	13 "	10 "	3 80	2¼ "	14 "	19 "	24 "	10 00
1¼ "	12 "	15 "	10 "	4 20	2¼ "	16 "	21 "	24 "	11 00
1¼ "	14 "	17 "	10 "	4 60	2¼ "	18 "	23 "	24 "	12 00
1½ "	5 "	8 "	12 "	3 50	2¼ "	20 "	25 "	24 "	13 25
1½ "	6 "	10 "	12 "	3 75	2¼ "	22 "	27 "	24 "	14 50
1½ "	8 "	12 "	12 "	4 25	2¼ "	24 "	29 "	24 "	15 75
1½ "	10 "	14 "	12 "	4 75	2½ "	6½ "	11 "	28 "	8 00
1½ "	12 "	16 "	12 "	5 25	2½ "	8 "	14 "	28 "	8 75
1½ "	14 "	18 "	12 "	6 00	2½ "	10 "	16 "	28 "	9 75
1½ "	16 "	20 "	12 "	6 75	2½ "	12 "	18 "	28 "	10 75
1¾ "	6 "	10 "	16 "	4 50	2½ "	14 "	20 "	28 "	12 00
1¾ "	8 "	12 "	16 "	5 00	2½ "	16 "	22 "	28 "	13 25
1¾ "	10 "	14 "	16 "	5 75	2½ "	18 "	24 "	28 "	14 50
1¾ "	12 "	16 "	16 "	6 25	2½ "	20 "	26 "	28 "	15 75
1¾ "	14 "	18 "	16 "	6 75	2½ "	22 "	28 "	28 "	17 00
1¾ "	16 "	20 "	16 "	7 50	2½ "	24 "	30 "	28 "	18 25
1¾ "	18 "	22 "	16 "	8 50	2½ "	28 "	34 "	28 "	22 00
2 "	5 "	9½ "	20 "	5 00	2½ "	32 "	38 "	28 "	26 00
2 "	6 "	10½ "	20 "	5 25	3 "	14 "	20 "	36 "	19 50
2 "	8 "	12½ "	20 "	6 00	3 "	16 "	22 "	36 "	20 75
2 "	10 "	14½ "	20 "	6 75	3 "	18 "	24 "	36 "	22 00
2 "	12 "	16½ "	20 "	7 50	3 "	20 "	26 "	36 "	23 25
2 "	14 "	18½ "	20 "	8 25	3 "	22 "	28 "	36 "	24 50
2 "	16 "	20½ "	20 "	9 25	3 "	24 "	30 "	35 "	25 75

Discount............

Levers will be sent only when ordered, and will be charged extra.

BELL BASE RATCHET JACK SCREWS.

This Jack has wrought iron screw, cast iron stand and cap, and steel ratchet, pawl and handle.

PRICE LIST.

Diameter of Screw.	Height Over All.	Lifting Capacity.	Price.	Diameter of Screw.	Height Over All.	Lifting Capacity.	Price.
2 in.	12 in.	24 Tons	$16 00	2½ in.	20 in.	32 Tons	$23 50
2 "	14 "	24 "	16 75	2½ "	22 "	32 "	24 50
2 "	16 "	24 "	17 50	2½ "	24 "	32 "	25 50
2 "	18 "	24 "	18 25	2½ "	26 "	32 "	26 50
2 "	20 "	24 "	19 00	2½ "	28 "	32 "	27 75
2 "	22 "	24 "	19 75	2½ "	30 "	32 "	29 00
2 "	24 "	24 "	20 50	2½ "	34 "	32 "	33 00
2 "	26 "	24 "	21 50	2½ "	36 "	32 "	35 00
2 "	28 "	24 "	22 50	2½ "	38 "	32 "	37 00
2 "	30 "	24 "	23 50	2¾ "	20 "	36 "	27 50
2¼ "	18 "	28 "	21 00	2¾ "	24 "	36 "	30 00
2¼ "	20 "	28 "	22 00	2¾ "	28 "	36 "	32 50
2¼ "	22 "	28 "	23 00	2¾ "	30 "	36 "	34 00
2¼ "	24 "	28 "	24 00	2¾ "	36 "	36 "	40 00
2¼ "	26 "	28 "	25 00	3 "	20 "	40 "	32 00
2¼ "	28 "	28 "	26 00	3 "	24 "	40 "	35 00
2¼ "	30 "	28 "	27 00	3 "	30 "	40 "	40 00
2½ "	18 "	32 "	22 50	3 "	36 "	40 "	48 00

Discount...........

TRIPOD RATCHET JACK SCREWS.

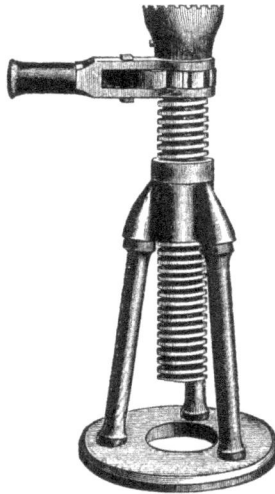

Our Tripod Jack Screws have wrought iron screws, legs and bases, brass nuts, and steel ratchets, pawls, and handles.

PRICE LIST.

Diam. of Screw.	Height Over All.	Lifting Capacity.	Price.	Diameter of Screw.	Height Over All.	Lifting Capacity.	Price.
2¼ in.	18 in.	28 Tons.	$50 00	2½ in.	36 in.	32 Tons.	$67 00
2¼ "	20 "	28 "	51 00	2¾ "	24 "	36 "	65 00
2¼ "	22 "	28 "	52 00	2¾ "	26 "	36 "	66 25
2¼ "	24 "	28 "	53 00	2¾ "	28 "	36 "	67 50
2½ "	18 "	32 "	56 50	2¾ "	30 "	36 "	68 75
2½ "	20 "	32 "	57 50	2¾ "	36 "	36 "	73 00
2½ "	22 "	32 "	58 50	3 "	24 "	40 "	70 00
2½ "	24 "	32 "	59 50	3 "	26 "	40 "	71 50
2½ "	26 "	32 "	60 75	3 "	28 "	40 "	73 00
2½ "	28 "	32 "	62 00	3 "	30 "	40 "	74 50
2½ "	30 "	32 "	63 25	3 "	36 "	40 "	79 00

Discount............

RATCHET CARRYING JACK SCREW.

This Jack has a steel base, brass nuts and wrought iron screws
and legs. The ratchets, pawls and handles are made of steel and
malleable iron.

Lifting Capacity ....36 tons.
Diameter of Lifting Screw .... $2\frac{3}{4}$ in.
   "   " Traverse Screw.......... $1\frac{5}{8}$ "
Length   " Lifting Screw.... ....18 "
   "   " Traverse Screw ....18 "
Height, over all.....................26 "

Price.. .. $140 00

Discount.............

TRACK JACK SCREW.

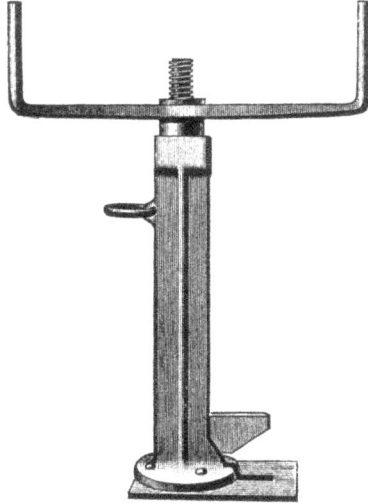

The screw, base and handle of this Jack are made of wrought iron. The lifter is welded to the screw.

Lifting capacity..................12 tons.
Diameter of screw............. $1\frac{1}{2}$ inches.
Length   " 18   "

Price... $13 00

Discount............

HOUSE-RAISING SCREWS.

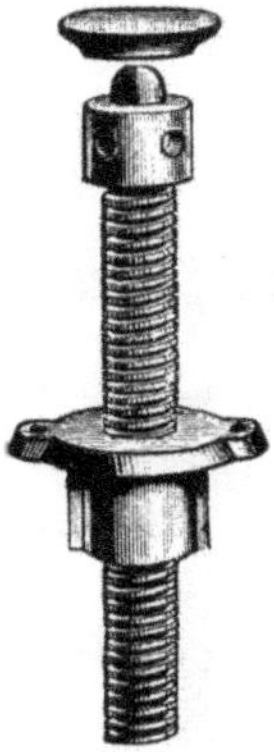

These Screws are made of wrought iron, with cast iron nuts and caps. The nuts are made to let into wood blocks.

LIFTING CAPACITY.

1¾-inch	Screw will raise		...............16 tons.
2 "	"	"	...............20 "
2¼ "	"	"	...............24 "
2½ "	"	"	...............28 "
2¾ "	"	"	...............32 "
3 "	"	"	...............36 "

PRICE LIST.

Diameter of Screws.	Height Over All.	Price.	Diameter of Screws.	Height Over All.	Price.	Diameter of Screws.	Height Over All.	Price.
1¾ in.	12 in.	$4 80	2 in.	24 in.	$ 8 40	2¼ in.	28 in.	$11 10
1¾ "	14 "	5 10	2 "	26 "	8 80	2½ "	12 "	8 50
1¾ "	16 "	5 40	2 "	28 "	9 20	2½ "	14 "	9 00
1¾ "	18 "	5 70	2¼ "	12 "	7 50	2½ "	16 "	9 60
1¾ "	20 "	6 00	2¼ "	14 "	7 90	2½ "	18 "	10 20
2 "	12 "	6 00	2¼ "	16 "	8 30	2½ "	20 "	10 80
2 "	14 "	6 40	2¼ "	18 "	8 80	2½ "	22 "	11 30
2 "	16 "	6 80	2¼ "	20 "	9 30	2½ "	24 "	11 90
2 "	18 "	7 20	2¼ "	22 "	9 80	2½ "	26 "	12 40
2 "	20 "	7 60	2¼ "	24 "	10 30	2½ "	28 "	12 90
2 "	22 "	8 00	2¼ "	26 "	10 70	2½ "	30 "	13 40

Discount...........

PRESS SCREWS,

FOR

PRESSING CIDER, WINE, TOBACCO, LARD, CLOTH, &c.

These Screws are made of wrought iron, with cast iron nuts and caps.

For lifting capacity see page 40.

PRICE LIST.

Diameter of Screws.	Height Over All.	Price.	Diameter of Screws.	Height Over All.	Price.	Diameter of Screws.	Height Over All.	Price.
2 in.	24 in.	$ 8 40	2½ in.	32 in.	$13 90	2¾ in.	48 in.	$24 80
2 "	26 "	8 80	2½ "	34 "	14 40	2¾ "	50 "	25 50
2 "	28 "	9 20	2½ "	36 "	14 90	2¾ "	52 "	26 20
2 "	30 "	9 60	2½ "	38 "	15 50	2¾ "	54 "	26 90
2 "	32 "	10 00	2½ "	40 "	16 00	2¾ "	56 "	27 60
2 "	34 "	10 40	2½ "	42 "	16 50	2¾ "	58 "	28 30
2 "	36 "	10 80	2½ "	44 "	17 10	2¾ "	60 "	29 00
2 "	38 "	11 20	2½ "	46 "	17 60	3 "	24 "	19 20
2 "	40 "	11 60	2½ "	48 "	18 10	3 "	26 "	20 00
2¼ "	24 "	10 30	2½ "	50 "	18 60	3 "	28 "	20 80
2¼ "	26 "	10 70	2½ "	52 "	19 10	3 "	30 "	21 60
2¼ "	28 "	11 10	2½ "	54 "	19 70	3 "	32 "	22 40
2¼ "	30 "	11 50	2½ "	56 "	20 30	3 "	34 "	23 20
2¼ "	32 "	11 90	2½ "	58 "	20 80	3 "	36 "	24 00
2¼ "	34 "	12 30	2½ "	60 "	21 30	3 "	38 "	24 80
2¼ "	36 "	12 70	2¾ "	24 "	16 40	3 "	40 "	25 60
2¼ "	38 "	13 20	2¾ "	26 "	17 10	3 "	42 "	26 40
2¼ "	40 "	13 60	2¾ "	28 "	17 80	3 "	44 "	27 20
2¼ "	42 "	14 10	2¾ "	30 "	18 50	3 "	46 "	28 00
2¼ "	44 "	14 60	2¾ "	32 "	19 20	3 "	48 "	28 80
2¼ "	46 "	15 00	2¾ "	34 "	19 90	3 "	50 "	29 60
2¼ "	48 "	15 40	2¾ "	36 "	20 60	3 "	52 "	30 40
2¼ "	50 "	16 00	2¾ "	38 "	21 30	3 "	54 "	31 20
2½ "	24 "	11 90	2¾ "	40 "	22 00	3 "	56 "	32 00
2½ "	26 "	12 40	2¾ "	42 "	22 70	3 "	58 "	32 80
2½ "	28 "	12 90	2¾ "	44 "	23 40	3 "	60 "	33 60
2½ "	30 "	13 40	2¾ "	46 "	24 10			

Discount...........

CHEESE PRESS SCREWS.

COMMON.   IMPROVED.

The Screws are made of wrought iron; other parts are cast iron.

PRICE LIST.

COMMON.			IMPROVED.		
Diameter of Screw.	Height Over All.	Price.	Diameter of Screw.	Height Over All.	Price.
1½ in.	16 in.	$ 7 00	1½ in.	16 in.	$ 7 70
1½ ''	18 ''	7 25	1½ ''	18 ''	7 90
1½ ''	20 ''	7 50	1½ ''	20 ''	8 10
1½ ''	22 ''	7 75	1½ ''	22 ''	8 35
1½ ''	24 ''	8 00	1½ ''	24 ''	8 60
1¾ ''	16 ''	8 00	1¾ ''	16 ''	8 90
1¾ ''	18 ''	8 40	1¾ ''	18 ''	9 20
1¾ ''	20 ''	8 80	1¾ ''	20 ''	9 50
1¾ ''	22 ''	9 20	1¾ ''	22 ''	9 80
1¾ ''	24 ''	9 60	1¾ ''	24 ''	10 10
1¾ ''	26 ''	10 00	1¾ ''	26 ''	10 40
1¾ ''	28 ''	10 40	1¾ ''	28 ''	10 75
1¾ ''	30 ''	10 80	1¾ ''	30 ''	11 10
2 ''	18 ''	10 75	2 ''	18 ''	11 20
2 ''	20 ''	11 25	2 ''	20 ''	11 65
2 ''	22 ''	11 75	2 ''	22 ''	12 10
2 ''	24 ''	12 25	2 ''	24 ''	12 55
2 ''	26 ''	12 75	2 ''	26 ''	13 00
2 ''	28 ''	13 25	2 ''	28 ''	13 45
2 ''	30 ''	13 75	2 ''	30 ''	13 90
2 ''	32 ''	14 25	2 ''	32 ''	14 25
2 ''	34 ''	14 75	2 ''	34 ''	14 70
2 ''	36 ''	15 25	2 ''	36 ''	15 15

Discount.... ....

WAGON JACK SCREWS.

These Jacks are made with wrought iron screws and cast iron stands.

PRICE LIST.

Diam. Screw, 1¼ in.	Height Stand, 12 in.	Lift. Capacity, 12 Tons	Price, $5 25
" " 1½ "	" " 14 "	" " 12 "	" 6 00
" " 1½ "	" " 16 "	" " 12 "	" 6 75
" " 1¾ "	" " 14 "	" " 16 "	" 6 75
" " 1¾ "	" " 16 "	" " 16 "	" 7 50

Discount...........

CAST IRON JACK SCREWS.

These Screws are cast with seamless threads, which makes them very smooth and uniform.

Diameter of Screw.	Height Over All.	Price.	Diameter of Screw.	Height Over All.	Price.	Diameter of Screw.	Height Over All.	Price.
3 in.	20 in.	$3 50	3 in.	26 in.	$4 25	3 in.	32 in.	$5 00
3 "	22 "	3 75	3 "	28 "	4 50	3 "	34 "	5 25
3 "	24 "	4 00	3 "	30 "	4 75	3 "	36 "	5 50

Discount...........

SAD IRONS.

SILVER POLISHED.

DIAMOND BRAND.

Our " DIAMOND BRAND " silver polished Sad Irons are superior
in quality and finish to any in the market. From the following
sizes we can make up any assortment the trade may require: 4, 5, 6,
7, 8 and 9 lbs. each, with sharp or round points. We pack them in
cases containing about 100 and 200 lbs. each, or in barrels of 600
to 800 lbs.

Price, silver polished .... per lb.
   "   nickel plated .... "

TAILOR IRONS.

FINELY POLISHED.

DIAMOND BRAND.

Can furnish Tailor Irons of the following weights: 12, 14, 16,
18, 21, 23, 25, 28 and 32 lbs. each, and large Sad Irons weighing 12,
15 and 26 lbs. each.

Price.................................... ... . per lb.

CAST IRON REVOLVING CLOTHES IRONS.

NO. 1.

A useful article, and one that sells readily.

Price..Per dozen, $15 00

NO. 2.

This Clothes Iron is strong, durable and ornamental; easily fitted to the top of a post, and by the use of it the annoyance and danger of clothes lines stretched across the yard or lawn is entirely avoided.

Price..Per dozen, $17 00

Discount ... .

GARDEN VASES AND RUSTIC HITCHING POSTS.

Finished in White.

Finished in Natural Colors.

FOR CONSERVATORIES, GARDENS, CEMETERIES, Etc.

PRICE.

	Complete.	Without Handles.	Without Base.
No. 1 Vase, 14 in. diameter, 23 in. high..........$3 50	$3 50	$3 00	$3 00
" 2 " 17 " " 28 " " 4 50	4 50	4 00	3 75
" 3 " 20 " " 33 " " 6 00	6 00	5 50	4 75
Rustic Hitching Post............................ 5 25	5 25		

Discount............

WIDE-AWAKE RABBITS.

NOT FRIGHTENED BY CATS OR DOGS.

These Rabbits are made of cast iron.

Finely finished in gray or white, they resemble the natural rabbit so precisely that dogs will jump and catch them at first sight ; but the same dog never attempts it a second time, but draws in his tail and runs off, hoping that no one noticed his attempt.

They are a handsome lawn ornament.

PRICE.

White finish........ .... Each, $1 25
Gray   " .... "   1 25

Discount...........

www.ingramcontent.com/pod-product-compliance
Lightning Source LLC
Chambersburg PA
CBHW030535210326
41597CB00014B/1156